The Indian or Cornish Game Fowl
Its Description, Characteristics, Origin, History and Breeding

by H.S. Babcock

with an introduction by Jackson Chambers

This work contains material that was originally published in 1891.

This publication is within the Public Domain.

*This edition is reprinted for educational purposes
and in accordance with all applicable Federal Laws.*

Introduction Copyright 2018 by Jackson Chambers

Self Reliance Books

Get more historic titles on animal and stock breeding, gardening and old fashioned skills by visiting us at:

http://selfreliancebooks.blogspot.com/

Disclaimer

This book was written in an age when cock-fighting was widely acceptable throughout society. In many places throughout the world, cock-fighting has been made illegal.

The material presented herein is intended to be strictly for educational purposes with the purpose of enlightening Game Fowl breeders about the history of their breed. Publication of the material is neither an endorsement, nor a criticism of its contents. This book is presented as part of large series of educational material on the history and raising of numerous chicken breeds for utility or exhibition purposes.

As the reader, please consider it your duty to become familiar with local, state, provincial and federal laws relating to the subject matter contained herein before attempting to utilize any of the information presented.

As the author, publisher and retailer cannot control how the reader utilizes the historical information presented in the pages herein, they hereby disclaim any liability to any party for any loss, damage, disruption or other liability that may be incurred by the reader's misuse of this material.

Introduction

I am pleased to present this third title in the "Game Fowl" series.

The work is in the Public Domain and is re-printed here in accordance with Federal Laws.

Though this work is a century old it contains much information on poultry that is still pertinent today.

As with all reprinted books of this age that are intended to perfectly reproduce the original edition, considerable pains and effort had to be undertaken to correct fading and sometimes outright damage to existing proofs of this title. At times, this task is quite monumental, requiring an almost total "rebuilding" of some pages from digital proofs of multiple copies. Despite this, imperfections still sometimes exist in the final proof and may detract from the visual appearance of the text.

I hope you enjoy reading this book as much as I enjoyed making it available to readers again.

Jackson Chambers

PREFATORY NOTE.

The wide-spread interest in the Indian Game fowl, the enormous and increasing demand for the breed, and the high prices which specimens represented as good readily command, are sufficient to tempt the cupidity of the unscrupulous and lead them to palm off, as genuine, fowls which are spurious Indian Games. If such fowls are inferior to the genuine stock the breed necessarily suffers in public estimation, and even if they were equally as good the public would be deceived and would gain an erroneous idea of what an Indian Game really was. To protect the public, by giving it a careful description of the breed, is one of the objects of this little book.

At this time the facts concerning the earliest importations of the Indian Game into the United States are sufficiently fresh in the minds of breeders to enable them to be stated accurately, but with each succeeding year these facts will grow more and more obscure, and in their places will appear myths more or less monstrous. There will never be a better time than now to record the simple, unvarnished facts, and preserve them for reference in the years to come. This work does this.

The subject of mating for the production of the finest specimens has, up to this time, received no treatment, although the articles on Indian Games which have appeared have not been few. The beginner with this breed desires first of all to know what to produce and then how to produce it. It is hoped that the suggestions contained in this work will be of real service to the beginner and not without value to the veteran in mating his fowls.

To protect the public, and by so doing to protect the interests of all breeders of genuine Indian Games, to record the beginnings of the history of the breed in this country, and to assist the breeder in mating his fowls so that the most perfect specimens may be produced are the leading purposes of this book, and if the author succeeds in accomplishing, in any degree, any or all of these, he will be more than compensated for the labor of its composition.

THE INDIAN GAME.

DESCRIPTION.

The Indian Game has a characteristic shape, peculiar to itself, yet resembling in some particulars that of other breeds. This shape is striking and prepossessing to those who like the general Game type, unprepossessing to those whose ideas of beauty have been moulded by breeds similar to the Cochin. It requires considerable modification of taste to make one who likes a very short legged, long and loosely feathered fowl, admire a fowl that shows some daylight beneath its body and whose plumage fits like a kid glove.

The head of the Indian Game is a compromise between the lean, long, thin type of the modern Exhibition Game, and the thick, heavy, beetle-browed type of the Malay. It is rather long, with a fair width to the skull and with eye brows which slightly overhang the eyes. In the Malay this overhanging skull gives a fierce and cruel expression to the fowl, but in the Indian Game it is just sufficient to give a look of courage and power. The strong, well curved beak, stout at its juncture with the head, adds to the powerful appearance of the fowl. The face is smooth and fine in texture, and the bare throat of the Exhibition Game is displaced by one dotted over with small feathers. The comb is an irregular pea, which often grows too large in some strains, and sometimes, like all combs, is weak at the base, causing it to lop over in an unattractive manner on the male. If the Standard, which may be adopted by the American Indian Game Club, permits the dubbing of combs, this defect, with others proceeding from its irregular character, will be removed. It is certain, also, that a dubbed cock has a more gamey look,—our ideas, of course having been moulded by the long established practice of dubbing Game fowls. A dubbed comb adds to the apparent length and leanness of the head, and therefore makes the fowl look keener and more wide awake. In the females the comb is small, and its triple character, especially in pullets, is sometimes difficult to discern. Yet the comb, even in such specimens, is very different from a true single comb. Of course it should be the ideal of breeders to produce a small and perfect pea comb, but, while such are represented in illustrations, and very rarely are produced in breeding, it will take some years of careful breeding with this end in view to produce a large percentage of the fowls with such combs. This, nevertheless, is to be said in favor of pea combs, however irregular and imperfect they are; if they are small their defects are inconspicuous and for practical purposes are of no moment. The best frost resisting comb may be a very small, irregular pea comb. The deaf ears, or ear lobes, should be small and the wattles very scanty.

The neck is rather more than of medium length. It is in fact a long neck if the fowl is compared with Asiatic breeds, but is medium in length if compared with the Malay. This term "medium," so often used in standards for breeds, needs to be interpreted with reference to the breed itself and not in reference to other breeds. There is no artificial standard with which to compare necks like the standards of a foot or a yard adopted in measuring. If the neck of an ideal Indian Game is described as medium it means that one of medium length for an Indian Game is better than either a very long or a very short one. This neck is carried very uprightly and is but slightly arched, the greater part of its curvature being near where it joins the head. And yet it is not an awkward, but, taken with the rest of the fowl, a graceful neck. The hackle is moderate in quantity, and shortness of feather is a desirable characteristic. The hackle should be just long enough to smooth the outlines and nicely cover the base of the neck, and should be perfectly free from twists.

The breast, owing to the upright carriage of the fowl, does not round out like that in some breeds. If the Indian Game carried itself more horizontally there would be a greater prominence to the forward sweep of the breast. Those who have given little thought to the subject imagine that a very pronounced forward sweep to the breast is evidence of great development of breast-meat, but what they think is meat is often only crop. The breast-meat is formed largely of the pectoral muscles which lie each side of the keel-bone, and is found, therefore, rather between the legs than in the forward sweep of the breast. But while the Indian Game does not have, and cannot have, owing to its military carriage, a pronounced forward sweep to the breast, it does have great width and fair depth, which indicate meatiness, and sufficient roundness to make the outline of the breast agreeable to the eye.

The body is very thick and compact, with great breadth at the shoulders. The back is flattish at the shoulders and ought to descend in a straight line to the tail, narrowing as it approaches the tail; but in many birds the upper line of the back is slightly arched, not so pronouncedly as in the Malay, and yet suggesting Malay characteristics. This is a trifling defect compared with the opposite one of being hollow-backed. The "concave sweep to the tail," which the breeder of Asiatic and American varieties so much admires, is foreign to the Indian Game, and such a bird would be very untypical. So far as this writer has observed, a hollow-backed Indian Game has not appeared at our exhibitions.

The wings are short for a Game fowl, the fronts carried prominently, the ends nicely rounded, and should be well tucked up. Of course in Indian Games, as in all other breeds, there are faulty wings; some loosely folded, some carried too low, some folded across the back. Between a wing carried too low and one too high, the preference would be in favor of the one carried too high—but just right is the ideal.

The thighs are of good length, but not so long as in the Malay, a breed with which the Indian Game is frequently compared, and are well provided with meat, giving the thigh a stout appearance. The shanks, especially in males, are very stout in bone, of good length but not extremely long, and the scales are smooth and neat. The description adopted by the Indian Game Club of England, in its Standard issued December 1st, 1886, and which has recently been readopted under the reorganization of this club, is worth quoting, as it sets forth the essential facts clearly. "The length of shank must be sufficient to give the bird a Gamey appearance, but in no case should it be as long as in the Malay, or in any way stilty." The best cocks which have been seen in this country have a good length of thigh and a medium length of shank, which gives them a tall but strong appearance. The toes are long, strong, straight, well spread, the hinder toes being set on low and pressing nearly flat upon the ground, ending with well shaped, strong nails.

The sexes vary, especially, of course, in the character of the tail. That of the male is preferred to be of medium length, carried at a slight elevation, making a broad angle with the line of the back, and furnished with narrow sickles and coverts. It is not so much the length of the tail as its shape and carriage, which gives character to the breed. It should be folded closely, not fanned out, and the longer sickles should curve over, giving an apparent drooping carriage. Narrowness of sickles and coverts gives a high-bred look, and is in keeping with the short, hard plumage demanded of this breed. The tail of the hen, owing to her slightly less erect carriage, is carried at a little greater elevation than that of the cock, but the angles formed by the back and the tail are the same for both. The tail of the hen is frequently considerably fanned out, but a closely folded tail is preferable, and such tails are now and then to be met.

The Indian Game is a large fowl. Cocks weigh from 8 to 10 pounds, and occasionally exceed this greater weight. The hens are much smaller than the males, and it is a good one that reaches 7 pounds, though greater weights are attained. Many hens weigh from $5\frac{1}{2}$ to 7 pounds,—a few are heavier. The difference between the weights of the males and the females is more marked in this breed than in most of our domestic fowls.

If these words of description convey the meaning that it was intended they should, they present to the mind of the reader a large fowl, powerful, broad, active, sprightly and vigorous, with an erect, military bearing, and a plumage short, hard and closely fitting. There are, however, to be noted one or two peculiarities which arise from the shortness of the plumage, and which, though not adding to the attractive appearance, ought not, in judging at least, to be regarded as defects. There is frequently on the wings a bare spot, and also on the lower part of the breast. These bare spots are natural characteristics and indicate a great economy in the matter of clothing, the fowl evidently being parsimonious in its expenditure for feathers, preferring to give more to the production of meat.

But, while in an Indian Game, shape is the first thing to be considered, for upon shape depends its great utility; color, also, has its demands and beauty its uses. The Indian Game is a richly plumaged bird. Its plumage is hard, lustrous, gleaming in the sun with the richest hues. The male is largely a black fowl, yet this black is relieved by other colors that add to his beauty. The plumage of the head is black. The hackle is black with a brown crimson shaft. A perfect hackle feather is a very beautiful feather, but in many cases this crimson spreads out beyond the shaft into the web of the feather, marring to some extent the beauty of the neck. But such males have their uses, as will hereafter appear. The breast is black, solid black in the best exhibition specimens, but as the females of the breed are laced, lace-breasted males occasionally are produced; and they, too, have their uses. The back is dark red and black, the latter color predominating, but the saddle shows more red than appears in the back proper. The wing is extremely handsome. Its bow is crimson and black like the back, its bar a rich metallic black, and its bay, formed by the edging of the secondaries, a rich bay. The tail is black, the sickles and coverts having a rich green lustre.

In both sexes the eyes are yellow or daw, and the shanks and toes a rich yellow or orange. Pale yellows sometimes appear, almost faded to white, and occasionally pullets show dusky on the front of the shanks, but as a breed it has as good yellow legs as any colored yellow-legged breed in existence.

The female is an exquisitely laced fowl, sometimes with a single, but oftener with two or more lacings or pencilings, one within the other. From the same pen both types of birds are bred. In England little if any preference is given to either type, but as the single-laced birds are generally the brighter in color, other things equal, they are more frequently the winners in English shows. In this country, if I can read aright the signs of the time, the drift of opinion is towards two or more lacings instead of one. The double-laced feather, as a feather, is certainly very beautiful, and if the lacings are sufficiently narrow and the ground color a rich golden bay, the effect on the fowl is exquisite. The tendency in double-laced birds is towards too wide a lacing, which almost obliterates the ground color and gives us, in appearance, a rich black fowl. The beauty of the female consists largely in the contrast between the ground color and the lacings, and this contrast lost the beauty is diminished, but not wholly destroyed, for the lustre of the black lacings is so great that even if it alone appears there is still considerable beauty left. My personal choice in color and markings for an Indian Game hen is a golden bay ground color, rich, clean and bright, with narrow, glossy black lacings, one within the other. But I think that a single-laced hen is handsomer than many of the double-laced, when the double-lacing is heavy and broad. Whether we adopt the English idea and allow double and single-laced birds to compete on an equality, or whether we adopt the double lacing as the American ideal, we ought to see to it that these sombre birds do not displace the much more beautiful bright colored ones with their contrasting hues.

A *dark* bay ground color, heavily laced with black, cannot compare in beauty with a ruddy gold ground color narrowly laced with black. The one affords almost no contrast between the colors, and what little exists is almost entirely concealed when the feathers are viewed as a whole, while the other affords a brilliant contrast in colors whether examined separately or seen in a mass. Too little attention has been paid to this subject in judging Indian Game females in this country.

CHARACTERISTICS.

While to the fancier the great beauty of the Indian Game is an entirely sufficient reason for its existence, to the practical poultryman there must be some further reason given. His first thought is of utility. He does not and he cannot afford to keep fowls simply to look at. They must be good for something, must be able to add to his income or he will have none of them.

The Indian Game, fortunately, is able to say something for itself upon this point. It is a splendid table fowl, the chief among chief breeds for this purpose. It has abundance of meat; the meat is fine grained, and beautiful in color; it is disposed most abundantly upon the choicest places; it is of the finest flavor. The skin and the shanks are yellow, and yellow is the favorite color in American markets. The breast is very full and round, and the breast, more than any other part of the dressed fowl, determines its attractiveness. The fowl grows to great weight, grows rapidly, and is plump in build at all times despite its length of leg. Few indeed are the fowls, if there are any, that can dress off so handsomely as an Indian Game, even when very young. It is, therefore, ready for killing at almost all stages of growth. It has received the highest endorsement abroad, and has there won many a prize in dressed poultry exhibits. It has the confidence and endorsement of those who are experimentally familiar with the breed in this country. I have myself killed pure bred Indian Games and Indian Games crossed on other breeds for my own table, and know that the reputation of the fowl is warranted by the facts.

For so good a table breed it is a good layer. We do not expect to find the most prolific laying and the best table qualities combined in one fowl. The prolificacy of the Leghorn and the table qualities of the Indian Game have not been and never will be united. But despite the antagonism which exists between great laying and great table qualities, the Indian Game has proved to be a very fair layer. Messrs. Sharp & Co. declare the fowl to be the best general purpose fowl in the world, and speak highly of its production of eggs. Mr. H. P. Clarke calls it a fine layer, and the best winter layer in the Game class. Mr. N. Reiner, who had kept other Games and Wyandottes, says it is the best laying fowl he ever had in his yards. And others have spoken in terms of praise of its laying qualities. I have always desired to avoid too great enthusiasm. I have been content to let the fowl stand preeminent as a table fowl, but I know that for a table breed it lays very well, indeed. The eggs are of good size, usually of a pale buff color, though sometimes quite dark, and are especially excellent for boiling or for use in fine cooking. Strains differ in the size of the egg produced. The hens I first imported laid very large eggs, while those of later importations have laid eggs somewhat smaller, yet of good size.

The fowl is a good setter, especially the lighter hens. I do not like very heavy hens for setters for they are apt to crush some of the eggs, especially if one happens to get some with thin shells under them. But the hens of this breed which weigh 5½ to 6 pounds make thoroughly reliable incubators, faithful brooders, and competent defenders of their young. Some of the hens break up easily when it is not desired to have them sit, while others are more persistent and take more time and patience to accomplish this result. But all will yield in reasonable time to the demands of the owner.

The Indian Game is a hardy fowl. The eggs hatch well and the chicks thrive. The following letter from a gentleman who had purchased Indian Game eggs, and which I have been permitted to inspect, is worth quoting in this connection:

"I thought you might wish to learn the result of the hatch from the sitting of Indian Game eggs (13) that you sent to me. * * * I got eleven (11) nice healthy chickens from the 13 eggs, the other two being unfertile. I call this an extra good hatch for eggs shipped by railroad, at least. *I have kept and bred poultry for nearly fifty years past, and have hatched chicks of many different breeds and varieties, but will say that* I HAVE NEVER HATCHED CHICKENS SO SMART AS THESE WHEN FIRST TAKEN FROM THE NEST. THEY RAN OUTSIDE THE COOPS IN THE SHORT GRASS BEFORE THEY WERE HALF A DAY OLD."

This letter is not a solitary example, but can be duplicated in essential facts by every breeder of Indian Games who has had and sold the genuine stock. And not only do they start well, but they keep the advantage that they possess in early life. It is discouraging to have a nice brood and then have them all die off; but an Indian Game chick hatched means a fair chance for an Indian Game fowl twelve months hence. I have tested the breed by the side of other breeds denominated hardy, and I have yet to see that the Indian Game has a superior in this very necessary quality. I do not hesitate to pronounce it a thoroughly hardy fowl.

The word Game, in its name, has led some to suppose that

it was a very pugnacious fowl, and that its keeping entailed almost endless trouble. Visions of scalped chickens and dead ones arose at the mention of the dread combination—Indian Game. But the facts are, that while it is not wanting in courage, that while the fowls are not easily cowed, it has been no more difficult to manage than the Plymouth Rock, and less difficult than the Leghorns I once used to keep. One can keep in one yard a hundred stags which have been reared together, and no trouble will ensue; but it is not advisable to introduce a strange cock into the yard with a lot of Indian Games, or, for that matter, with a lot of meek-eyed Asiatics. There is danger of a general melee under such circumstances. No one, who has the courage to keep Cochins, need be deterred from keeping Indian Games because of their pugnacity. The Indian Game is a courageous but naturally quiet, peaceable fowl, agreeable to manage, if it is managed properly.

Being a heavy fowl, and having a short wing, it is not difficult to confine this breed. A fence which will confine a Plymouth Rock is usually quite sufficient to keep an Indian Game within bounds. Unlike most of the Game family, the Indian is not a high-flyer.

ORIGIN.

The origin of the Indian Game has been robbed of much of its mystery, if we can believe the statement of Mr. John Frayn, as published in the *Fanciers' Gazette* of London, England. Under the title "CORNISH INDIAN GAME," this paper sometime ago published the following:

"No variety of poultry has made greater progress of late years than the Indian Game, whether at home or abroad. It is not so very long ago ere this breed began to make its appearance in our shows, and to claim a place in poultry exhibitions. Many of those whose purview was not restricted to feather considerations, knew that in the far southwest there was a breed outwardly handsome and inwardly valuable, the result of careful selection super-imposed upon a basis of high economic merits. Locally it was regarded as one of the finest breeds for table purposes, but not until the fancy as such got hold of the Indian or Cornish Game did it win that place in the wider arena of life, which is so well deserved. But when once the much maligned fanciers recognized its value, then there was opened out a career of influence which, we venture to think, is only in its adolescent stage even now. Possibly, nay, probably, the favor with which it is now regarded is due as much to its external beauties, which are great, as to its merits, from the economic point of view, for which too much can hardly be said; but it is here where the benefits of the fancy come in, for the great popularity achieved by it has been the means of spreading it far and wide, giving it a niche in the hall of fame peculiarly its own.

"At St. Stephens, on the north of the lovely valley wherein the town of Launceston clambers upon the southern slopes, and within sight of the famous old castle whose history links past and present, and whose circular ruin presents features in the architecture of defensive strength, is the home of John Frayn. That gentleman's rubicund physiognomy accompanies these notes, and may be taken as a representative type of the true fancier, one who has thriven with the breed, for its success has been his, and his efforts on behalf of the Indian Game have met with the reward they so well deserve. Mr. Frayn has been a breeder of Indian Games ever since he was a boy, following in the footsteps of his father, and his experience of more than thirty years makes him one of the oldest fanciers of this variety living. In his early days it was a short, squabby fowl, ginger in color. The great change which has come over the breed is due to crossing with the rich-plumaged black Indian Game, and not to Malay blood, as many aver. About sixteen years ago Mr. Frayn began to pay attention to the marking of the females. He owned one famous hen, which was ten years old when she died, and to her influence is attributable the great improvement in the direction named, her blood being paramount even to-day. So far as Mr. Frayn is aware, the origin of the breed is due to direct importation. His father secured the first specimens from a farm in the west, which were of the type already named."

Even if we accept the above as the exact truth, which I am inclined to do, for I find that there is a tendency in the breed to now and then produce a black chicken, or an off-colored one suggestive of the old ginger color, a tendency not at all marked but which really does exist, the further questions remain to be answered—what was this stout, squabby, ginger-colored fowl? Whence came it? Whence came the so called black Indian Games? To these questions it may be difficult to give any satisfactory reply. I have seen a fowl, imported from India, of which the females are light brown and the males black red, which, outside of color, answers to the description given above—"short," "squabby." These fowls have the general characteristics, otherwise, of Indian Games. But it is altogether probable, that even if Mr. John Frayn's statement is to be taken literally, there still exists a relationship between the Malay and Indian Game. It may not be and probably is not one of parent and child, but both may be descendants of a common Indian stock developed in different directions—the Malay into the stilty, cruel looking fowl of to-day, the Indian into the beautifully marked and elegantly proportioned fowl that Mr. Sharpe has called "The bird of destiny." It seems to me that some such explanation as the above will account for the resemblances and the differences in and between these two birds. But if it be of composite origin, and if the Malay be one of the progenitors of the Indian Game, this fact need not disturb us. If the Indian Game is a made breed it has been well made, for cross it upon almost any variety of fowl and the progeny will show more characteristics of the Indian Game than it will of the other breed represented in the cross. The power to perpetuate its characteristics is the supreme test of the thoroughbred character of any breed, and that power the Indian Game possesses in a highly marked degree. Whatever its origin there can be no reasonable doubt but that the Indian Game is a strictly thoroughbred fowl.

HISTORY IN ENGLAND.

Little remains to be added to the quotation made from the *Fanciers' Gazette* concerning the history of this fowl in England. It had long been bred for its economic properties in Cornwall, but had cut no figure in the great English exhibitions. Finally the fancier, to whom the practical poultryman is indebted for the useful breeds he possesses in a larger measure than he is willing to acknowledge, took hold of this local breed and gave it a world-wide fame. In 1886 the Indian Game Club was formed in England, if I am not mistaken, and at a general meeting of the club held at the public hall in Devenport, on the first day of December, 1886, a Standard was prepared and adopted for the breed. Julius G. Mosenthal was the first Hon. Secretary. From this date the Indian Game leaped from comparative obscurity into the full blaze of public popularity, though, owing to the absence of Mr. Mosenthal on the Continent, the club after a time lapsed into a state of "innocuous desuetude." In the first part of 1891 this club was reorganized with John Frayn as President, his brother James as one of the Vice-Presidents, and Mr. Geo. T. Whitfield as the Hon. Secretary. The Standard framed in 1886 was readopted in 1891, and is now the recognized Standard of the breed in England.

HISTORY IN THE UNITED STATES.

In 1887, attracted by the favorable notices of the English poultry press, Mr. H. S. Babcock imported from J. G. Mosenthal, the Hon. Secretary of the Indian Game Club, both eggs and fowls. The eggs were received early in the season, but the fowls, owing to the absence of Mr. Mosenthal upon the Continent for two months, were delayed in their arrival. That same year, Mr. H. P. Clarke, of Indiana, imported a trio of Indian Games, but from what breeder I have been unable to learn. From facts in my possession, I think that the following statements are the truth in respect to the earliest introduction of the Indian Game into the United States.

1. Mr. H. S. Babcock was the first American fancier to call public attention to the merits of this fowl, was the first to import Indian Game eggs, was the first to place an order for Indian Game fowls with an English breeder, and was the first, or one of the first two, to receive such fowls. It is possible that Mr. H. P. Clarke, owing to the delay in filling Mr. Babcock's order, was the first person to receive the fowls, though the earliest date mentioned by him is considerably later than the reception of eggs and the ordering of fowls by Mr. Babcock.

2. Mr. H. P. Clarke was unquestionably the first American breeder to exhibit, in this country, Indian Game fowls, and, as above stated, is possibly the first who actually received such fowls from England. Indian Games were exhibited by him first at the Indiana State Fair in September, 1887, and next at the National Poultry Show, in the City of Indianapolis, in January, 1888.

3. In January, 1889, Mr. George T. Whitfield, of Market Drayton, England, sent a pen of Indian Games to the Buffalo, N. Y., International Exhibition. This pen of fowls was subsequently purchased by Sharpe & Co and the cock named Agitator, and Messrs. Sharpe became identified with the breed in this country. That same year Mr. C. A. Bowman imported stock from Frayn, and Mr. Babcock increased his stock by importing from J. Penfold Field, who a few years previous had begun with the stock of John Frayn. The next year, 1890, saw a wonderful increase in the number of importers and importations. Messrs. Sharpe made a very large importation from various leading breeders, Mr. Bowman introduced more Frayn birds, Mr. Babcock imported from James Frayn and others, Mr.

Aug. D. Arnold imported from James Frayn and Abbott Bros., and Mr. P. A. Webster and others imported from Abbott Bros. and other large dealers. The number of breeders, not only of those who have made direct importations, but of those who have bought imported or American bred stock, has steadily increased until now this country is practically independent of England for its Indian Game stock. It has representatives of the best English strains thoroughly acclimated, and therefore better suited to the needs of American poultrymen. It is probable that American bred birds will now have the call and that importations will be more sparingly made, and only by those who wish new blood. And yet even this will be rarely needed, for American breeders can furnish all the new blood necessary for many years to come.

From the start the sale of Indian Game eggs and fowls has been phenomenal. All the leading breeders have been crowded with orders. The simple fact is the demand has greatly exceeded the supply, and instead of soliciting orders one had to drive away buyers. And the end is not yet, for the demand continues strong and the supply weak; there are two buyers for every bird that is for sale. The comparatively short time that the Indian Game has been in this country has been long enough to transform an obscure breed into one of the most popular fowls in the United States. It has enjoyed and is still enjoying an almost unprecedented boom. It took from the start and it has been taking ever since. It has, of course, awakened opposition, and the enemies of the breed are numerous. By misrepresentation due, it is charitable to suppose, to ignorance of its merits, they have done and are doing all that perverted and misdirected human ingenuity can do to destroy its popularity. Their opposition thus far has but added fuel to the flames, and, to vary the figure, the stronger the wind of opposition has blown, the higher the tide of popularity has risen. The Indian Game could say in the words of Ceasar, slightly altered, "I came, I was seen, I conquered."

At the New York show, in 1890, the American Indian Game Club was organized, with Mr. H. S. Babcock as President, and Mr. O. K. Sharp as Secretary and Treasurer. At the annual meeting in New York, in 1891, the President and Secretary were re-elected, and the latter reported a list of over sixty members with encouraging prospects that the number would soon be one hundred or more. It is already one of the strongest specialty clubs in the country, and if it continues to grow as rapidly it will be but a short time before it will lead all the others. It has not at this writing adopted a Standard of its own, but such a Standard is preparing by a committee of the club, and before these words are printed, may be adopted. The club has temporarily adopted the English Standard, which will be in due time supplanted by the American.

STRAINS.

There are, strictly speaking, no true American strains of Indian Games, though we hear of the Babcock strain, the Clarke strain, the Agitator strain, and the like.

The Babcock strain, so called, is composed of blood elements from Mosenthal, Frayn, Field and Percival. The Clarke strain, so called, I am unable to give all the components of. The Agitator strain, so called, was built upon the Whitfield birds, but has been added to by subsequent importations purchased of various English breeders. Mr. Bowman has kept strictly to Frayn stock. The Field and the Whitfield stock trace directly back to the Frayns, John and James, as probably do many other so called English strains, so that really there are but few strains in England and none, strictly speaking, in this country. But the older breeders of Indian Games are gradually breeding to type peculiarities and there are incipient strains in this country which in time may become established. It is already possible for one who has a keen eye, well trained in observing nice points, to shrewdly guess from what yards certain birds come. But until, with enough yards to prevent dangerous inbreeding, American breeders stick to their own fowls and cease importing yearly from abroad, there will be no established American strains, and the so called strains will simply mean birds bred by Mr. A, or Mr. B, or Mr. C. It takes time for a skillful breeder to establish a strain, and an unskillful one never can do it. Sufficient time has not yet elapsed for any American breeder to establish a strain.

MATING.

In the early days of breeding Indian Games the subject of mating received little attention. Fowls were bought in pairs, trios or pens, and as they came they were breed. But as time has gone by and results have been noted, better matings became possible, matings which will give more satisfactory results. The following advice upon mating is not to be taken as final, but as the best which observation has thus far suggested. In the uncertainly as to what the American Standard will be, it will be necessary to assume some sort of an ideal bird to be produced, and the following matings are for the production of birds similar to those described at the beginning of this work, with distinct double lacing in the females.

THE SINGLE MATING.

Admirable birds can be bred from a single mating, but it is necessary to employ two types of females. In this, as in all matings, select birds which are typical in shape, with prominent wing-fronts, good width of shoulder and breast and strong boned. The male bird, cock or cockerel, should be absolutely free from white in plumage, if he can be had, but if not, of course the less white the better. White is a common but serious defect in this breed, and should, so far as possible, be avoided in the matings. His hackle should be solid black, except the shafts of the feathers, breast solid black, and in fact the bird should be throughout such a one as would be demanded for exhibition purposes. With him should be mated one or more double laced femalas, the ground color a dark bay, and the lacings, especially the outer one, very broad and intensely black. Such females will produce admirable cockerels. There should also be with him, one or more females, golden bay in ground color, with very narrow, yet distinctly defined double lacings. These females look very light in the pen, but mated to such a male produce excellently marked and richly colored pullets. From them will probably be bred some laced breasted cockerels.

COCKEREL MATING.

Mate an exhibition cockerel showing a great preponderance of black in the plumage with very dark, heavily laced females. It is unimportant, so far as the cockerels are concerned for exhibition, whether these females are double or single laced, but if they are to be used in breeding and the double-laced bird is the one favored in our exhibitions, it is preferable that they should be double laced.

PULLET MATING.

Mate a cockerel showing more red in the hackle than is desired for exhibition purposes—one where the red has crept out beyond the shaft into the web of the feather,—with a bright crimson red saddle having little black intermixed, and with a laced breast—the centers penciled if possible,—the fowl otherwise being an exhibition bird, with females having a rich golden bay ground color and distinct double lacings of medium width. From such a mating will be produced beautifully-laced bright colored pullets, but most of the cockerels will be unfit for exhibition purposes, being too light in color and having laced breasts. Such a male, however, mated to very dark, heavily laced females, will produce some good cockerels and many good pullets. Such a mating is really a very good one for those who have room for but one pen, though scarcely equal to the first mating mentioned.

Another and very satisfactory mating is to select a male bird showing but little lacing on the breast, a good hackle, free from red in the web of the feather, but having instead of the dark crimson a brighter crimson where the red should appear, and having a bright, rather light wing bay, and mate him to such pullets or hens as one would wish for exhibition, golden bay in ground color, sharply laced with two or more medium lacings. The cockerels may be a trifle light in color, but will be likely to make fairly good show birds, and the pullets will be largely of the desired hue and marking.

While the above are but suggestions, and made somewhat in the dark on account of the uncertainty of exactly what the American Standard will be, yet it is hoped they will prove of value to the breeders of Indian Games, and will assist the beginner, at least, in producing specimens fit to compete under any Standard that will be likely to be issued by the American Indian Game Club.

CONCLUSION.

Although the Indian Game is now enjoying a great boom, and although the present indications are that this boom will continue for some years to come, it would be folly for any one to believe that it would never end. Now an Indian Game is in demand solely because it is an Indian Game, but the time is approaching when the demand will be for only first-class birds—fine birds for exhibition or good ones for breeding. The winnowing process will begin and the wheat will be separated from the chaff. The best birds will be demanded, high prices will be realized, but many which once would have been sold for breeding will fitly furnish the table of some epicure. It is, therefore, the part of wisdom to prepare for this time, by carefully selecting breeders and mating them so as to produce the finest possible specimens. The breeder who can best do this will be the one who will obtain the highest prices and receive the largest demand for his stock. The future of the breed is safe, for it has so many and so great merits, for use and for beauty, that it cannot fail of maintaining a position in the front rank of breeds. To boom the breed it is only necessary to follow the advice of Ex-President Cleveland, "Tell the truth," for the truth comprehends so much of praise that it sounds to the uninitiated like an eulogy. A breed of which this can be truthfully said has no uncertain future before it, and this can be truthfully said of the useful and beautiful Indian Game.

THE END.

www.ingramcontent.com/pod-product-compliance
Lightning Source LLC
Chambersburg PA
CBHW062236220526
45471CB00009B/3507